Koalafied

The story of the koala species

By Benjamin Minch

Cover Illustrated by Khang Do

Table Of Contents

Chapter One: The Koala Problem   5

Chapter Two: You are What You Eat   17

Chapter Three: Behave Like a Koala   37

Chapter Four: They're All the Same   55

Chapter Five: Koala Cousins and Fake Trees   61

Chapter Six: The Law Breaking Koala   79

Chapter Seven: The Problem Down Under   85

Chapter Eight: Dingoes on the Decline   95

Chapter Nine: The Future of Australian Biodiversity   111

Chapter Ten: A Case for Life   137

Chapter One:
The Koala Problem

By no means are they a keystone. By no means are they carnivorous hunters that keep populations of smaller herbivores down. But they are important nonetheless: and they are on the decline.

On a recent trip to my local zoo, the San Diego Zoo in San Diego, California, I couldn't help but stop by the Koala exhibit. I was one of the only ones there. It seemed as though nobody cared for, or about them. I mean, they are perfectly justified in doing so because Koalas aren't the most interesting animals. Most of the Koalas had rugged fur and were either scratching themselves or sleeping. Many of the branches in their habitat were gnawed and the place looked like a wreck. But I stood there, admiring my favorite animal. Just then, one of them woke up, and with it a sense of playfulness and a whole new personality. It was as though the animal that was asleep wasn't the same anymore. This Koala I remember so vividly was playing with a red ball on the ground. It was almost as though he wasn't a Koala at all. Its face was pale and white, and I could tell that it was getting old. But it still played. More than any Koala I had ever seen (because the only other ones I had seen had been asleep). It was only for a short while, but this act of oddness gave hope to me, and the entirety of the Koala population.

Many know that Koalas are devout sleepers, or have a slim diet, but not many know how vitally threatened the species is. With the creation of zoos, I feel that many people have adopted this view about multiple species. Zoos make the oblivious, average visitor feel at home. The visitor doesn't need to ask hard questions because everything is written on information cards, or need to wonder about animal safety because the animals in the zoo seem to be accurately representing animals in the wild. Nobody will question if an animal is receiving the proper care because all they see is the surface: a pretty cage with a pretty animal that has food and shelter. "What more could an animal possibly need?" they wonder. Many think of zoos as animal shelters and that they are saving animals, when in reality they are constructs with the sole purpose of making money.

According to an organization devoted to stopping animal cruelty, The Last Chance for Animals,:

"Animals in zoos are forced to live in artificial, stressful, and downright boring conditions. Removed from their natural habitats and social structures, they are confined to small, restrictive environments that deprive them of mental and physical stimulation. While zoos claim to provide conservation, education, and entertainment, their primary goal is to sustain public support in order to increase profits."

Animals in zoos also develop symptoms, which have been classified as "Zoochosis" by the Last Chance for Animals organization. These symptoms include

Bar biting

Coprophagia (Consuming and playing with excrement)

Self-mutilation

Circling

Rocking

Swaying

Pacing

Rolling, twisting, nodding of the neck or head

Vomiting

Frequent licking

Excessive grooming.

Many of these behaviors are unnatural to animals, but quite sadly, they get bored in a cage. They are living artificial lives that are boring, painful and unrepresentative of the species as a whole. The Koalas I had seen had displayed many of these behaviors, so I wondered if they had been victim to Zoochosis as well.

In 2010, a study was done to compare the well being of Koalas in the wild vs. at the zoo.

Veterinarians Brian and Geoff set out to St. Bees Island off the east coast of Australia in order to study the Koalas there and compare them to the Koalas of the San Diego Zoo. All of the Koalas they had studied on St. Bees island had been healthy, some of which were old females who still were giving birth to new young. Koalas at the zoo appeared to be in the same physical condition as those on St. Bees island, but this is before recent studies about animal boredom: a problem that plagues many zoo animals and has probably affected Koalas as well.

A recent research project done by the University of Guelph studied boredom in caged animals. Half of the animals studied, hamsters, were put into bare cages with nothing to do, while the other half were put into cages with many different toys and wheels to play with. The animals in the bare cages were more likely to eat more treats than those in the other cages and when presented with a stimulus, they would respond almost immediately. Still, little is known about the concept of boredom or what its effect on animals is, but one thing is for certain: it isn't a good thing to have.

Although the zoos do cause boredom for the Koala, they serve as a key factor in keeping the Koala species alive. If they are in the zoo, the Koala isn't at risk of the greatest habitat loss in the history of Australia: Urbanization.

Koalas exposed to urbanization, a growing problem of overpopulation and industrial expansion, are much more likely to die of deadly chemicals and diseases such as Chlamydia and dirty tail disease. Chlamydia is a very common strain of bacteria found most commonly in Koalas living in urban areas such as Brisbane. Oddly however, many Koalas on St. Bees island also have the disease, but are unaffected. Why is this?

Research done by Brian and Geoff on their Koala expedition has shown that the strains found in both locations are genetically different. The one found in the urban location was more deadly because it had absorbed toxins caused by pollution and smog, while the strain on St. Bees Island was less deadly.

Chlamydia isn't the only thing threatening Koalas with the increase of urbanization. Eucalyptus forests are being torn down, and roads and cars are exposing Koalas to potential dangers such as getting hit by a car. But this isn't even the largest threat to Koalas at this time. Urbanization is a growing problem, but Koalas are for the most part under the protection of the government. The biggest modern threat to Koalas as well as various other species around the world is Global Warming.

Not only are Eucalyptus forests being cut down, they are also dying from a drought caused by the increasingly hot temperatures and little rainfall over eastern Australia. The increasing temperatures caused by global warming cause dry grass, which in turn, causes more brush fires that destroy the

Eucalyptus trees. This intense heat also drains water from the Eucalyptus

This poses a real problem for the Koala population because Koalas don't actually drink water. They get their water almost entirely from Eucalyptus leaves. If Eucalyptus leaves dry out, Koalas will have no water, and will die...Or will they?

March 31, 2017: Artificial water stations were placed in Gunnedah, Australia by the University of Sydney, a recent project they had been working on. To their surprise, Koalas started coming down from their cozy position in the trees to drink out of these water stations. It was a massive observation showing how the Koalas are adapting to the climate change, that they are capable of survival through the selective pressure of climate change. Many believe that this learned behavior is based on previous research in which Koalas will reject Eucalyptus leaves that have less than 55% water content. Previous to putting in the water stations, the Koala population in Gunnedah dropped severely in 2009 because of a heat wave.

This new learned behavior is very interesting because the word Koala literally means "no drink" because they don't drink water. Can learned behaviors change a species completely? Hopefully through this book, this question and many others like it will be answered. On top of this, the radical behavior change also proves the ability for the Koala to adapt to survival. Some may see this as an example of natural selection at work: but why then, don't Koalas evolve to eat a more abundant and nutritious food source? The number one cause of death in Koalas is starvation. Why haven't they evolved?

Many other questions can be asked about such an obscure phenomenon. Could the Koalas be under stress from the heat? This event in itself has caused me to conduct months of research, gathering all the knowledge I know of Biology, Koalas and Ecology and finding out why this would happen and what it means for the continent as a whole.

Let's look at the facts, 65% of all Eucalyptus forests have been cleared in Australia. Koala populations are on the decline by 60% over the past years due to urbanization, lack of food and climate change. There has also been an 80% decline in the moisture of Eucalyptus, causing more fires and less water for the Koalas to drink. On top of all this, Koalas are a K selected species, meaning that they have a slow breeding time and don't populate very fast: only about one baby per year. All these facts may seem hopeless, like there is nothing that we can possibly do to reverse this process. I am here to assure you there is a way. I wouldn't be writing a book if there weren't.

The purpose of this novel isn't to demoralize humanity, or look down on what we have done. It is to provide solutions, to see clearly the path to solving the problem of Koala population decline. But most of all, I am writing to expose the conditions that real Koalas are really facing. Not an "artificial" Koala at the zoo that appears to be fine. Not some information sheet showing you how much Koalas sleep, or eat. I am here to show you how much wild Koalas need us, now more than ever. You will no longer be an standard "zoo viewer" after reading this. You will have all the tools for change. All the knowledge and facts to defend your case... It is just a matter of if you want to be the change you wish to see.

Chapter Two:
You Are What You Eat

The common phrase "you are what you eat" is only partly true. While food does make up a great portion of who we are, it is in unison with genetics, more particularly: epigenetics. Epigenetics is the study "above the genome", which studies how different genes can affect how your metabolism works, or your addictions or habits. Epigenetics works by the removal and addition of certain Methyl groups of molecules that can bind to histone proteins on dna to either unwind it or make it coil. DNA can only be expressed if it is open because it is the only way that the proteins can transcribe, replicate and translate the DNA strand. If DNA is closed or coiled up, it cannot be replicated or transcribed into RNA. This form of genetics can be responsible for multiple diseases, addictions and habits. For example, drug addiction can be an epigenetic trait. If your parents were addicted to a certain drug, certain methyl groups of their genetics were removed and expressed differently. These genes are then passed on through inheritance to you, who now is more likely to become addicted to drugs based on your parents.

In rats, epigenetics plays a key role in determining the survivability of the rat. When rat babies are newly born, they can be licked or neglected by their parent. Licking releases the Methyl groups attached to the uncoiled DNA. Rats who are licked will have a lower level of stress as they grow older and will be more relaxed and feel loved. Rats who aren't licked at birth will have a lot of stress and fear in their life, causing them to live for a shorter amount of years. Stress however, can be beneficial for the rat if he is in an environment with a lot of predators and needs to be alert of oncoming dangers. It is ultimately up to the mother of the rat to determine the baby's fate.

I believe that this situation with the rats is the same as with Koalas. The licking of rats is very similar to the eating of eucalyptus by the Koala. The habit of

Koalas eating eucalyptus probably came from an epigenetic trait that was beneficial in the past. But in order to understand this concept more, we must look at the anatomy and physiology of the eucalyptus tree.

Eucalyptus

The eucalyptus tree is one of the tallest plants on the planet. With over 700 different species, most of them are evergreen, meaning they stay green all year round. This is very advantageous for a food source because it means it is available and edible all year. This is probably one of the reasons that the Koala has chosen to eat this as a food source. Koalas are low mobility animals, so it seems reasonable that they would choose a very abundant food source such as eucalyptus.

Recently however, this food source has become less abundant. Eucalyptus is used for wood purposes, and is also used to make paper and oils.

The mass movement towards urbanization has also caused deforestation to occur in order to build more homes. Not only are eucalyptus forests in decline, the eucalyptus is also growing smaller. Based on a study done by the Royal Botanic Garden, a government agency, eucalyptus trees grown after 1991 are mostly in the 0-.5 meter range when their normal height should be around 1 meter. This decline in the size of eucalyptus means that there is also a decline in food for the Koala.

For this very reason, the Australian Koala Foundation has started a movement to protect and restore the eucalyptus forests in Australia because they know that without trees, there is no Koalas.

But shortage of food isn't the only problem Koalas are facing; it is also the content of the food. Eucalyptus is the only water source for the Koala and they depend fully on it for that nutrient. The water content of eucalyptus however, has declined by 80%, meaning Koalas will be unable to acquire sufficient nutrients. The eucalyptus leaves are also very lacking in nutrients and calories. This plant has one of the lowest ranking calorie count of all plants. This is why Koalas must sleep for so long: they don't get much energy from their food source. In spite of this low energy, Koalas must digest and extract every last from the leaf. They do this through the caecum, an organ that digests fiber. Humans have a caecum as well, but ours is very small and irrelevant. The caecum of the Koala is 200cm long, proving LaMarck's theory that organs that are most used will remain and be strengthened in an organism. The caecum contains millions of bacteria that are designed to break down fiber. These aerobic bacteria survive from the oxygen gained from the plants. These bacteria can perform cellular respiration because they are aerobic, meaning they can break down compounds faster. Even after this process of breaking down the fiber, only 25% of the fiber is absorbed. This tiring and energy consuming process is another reason that the Koala must sleep so long.

On top of the low nutrient count of eucalyptus, the plant is also highly toxic. This means that more energy is required to break down the compound. The toxins however, are actually advantageous to the Koala. After eating eucalyptus all of their life and storing it in their stomach, the Koala pretty much becomes immune to all toxins, including those released in air pollution and those from urbanization. This is a good thing for the species because they are more resilient to the changing urban culture.

By now you have probably figured out why Koalas sleep so long and barely move at all. They eat so little calories, that moving long distances would cause them to use up their little energy, and they sleep so long for the same reason. This immobility causes many problems though. If a Koala's food source in an area is depleted, it will probably be unable to make it to another source because it cannot move very far. This is happening more often because of the smaller size in leaves as a result of hotter temperatures from global warming. On top of this, Koalas are less capable to adapt to deforestation of an area than any other animal in that area. By this time, it is a strange phenomenon that Koalas haven't been naturally selected straight out of Australia, one of the most dangerous places on the planet.

Hungry Koalas

All of these issues with food are what causes the number one cause of death for Koalas: starvation. Starvation is responsible for 50% of Koala deaths in Australia, a number so significant that it was enough to change their lifespan. Koalas aren't victimized by any other predator besides their food source, meaning they should live longer than they are currently. The heavy digestion process required by the Koala to break down eucalyptus causes their teeth to wear out very fast. This causes them to be unable to extract the nutrients required from the eucalyptus plant and will ultimately cause death. This however, is only one reason of starvation. Another way that Koalas die of starvation is the destruction of eucalyptus forests. Every time a forest with Koalas living in it is cut down, the Koalas will be unable to migrate to another forest for food and will most likely die of starvation.

Starvation however, is completely preventable. Will a human starve if there is no more candy? The answer is no. Humans as well as many other organisms and animals have a wide range of diet. They can eat almost any food as long as it meets two categories: it has to be present, and it has to contain nutrients. These are the two things necessary for food to be a sustainable part of an animal's diet. Take polar bears as an example. Polar bears hunt almost exclusively seal pups, as they are present and provide the nutrients that are necessary for the survival of the polar bear. Recently however, global warming and the melting of the ice caps has caused the numbers of seals to decline. These seals that were once abundant within the area of the polar bear have migrated elsewhere because of the climate. This posed a challenge for the polar bears that is similar to that of the koala. Do they sit there and starve to death?... Or do they find a new source of food? The polar bears have chosen the second route, which is the basis of adapting to the changing world around us. In 2013, a paper was written for the journal of *Polar Ecology* that reported seeing polar bears for the first time attacking and eating snow geese during the spring mating season. A second paper published to the *Ecology and Evolution* journal reported that polar bears were seen pursuing and eating caribou more than attacking seals. The final paper published on this topic in 2013 to the *BMC Ecology* journal showed polar bears to be eating plants as well as geese and caribou, and have also started moving less. These polar bears are showing incredible resilience to the challenge of climate change in front of them: something that should've killed them a while ago. If polar bears were unable to adapt to the lack of food, they would have surely died out and become an animal of the past. This resilience is what drives biodiversity to survive, but it hasn't yet come upon the koala.

So what does resilience look like for the koala? Eucalyptus forests are in decline, not just in number, but in nutrient content and water as well. The 80% water decrease in the leaves has already sent koalas looking for other sources of water such as the drinking fountains set up by the University of Sydney. Koalas have already started to succumb to the negative effects of a lacking food source, which is starvation. As eucalyptus, like the seals, decline, koalas must find a new food source. Here are some potential options…

Banksias: Banksias is a native Australian plant that is abundant in southeastern Australia, right where the koala lives. This plant is well protected by the government of Australia. On top of this, the plant has no conventional purpose like the eucalyptus does such as paper, oil and lumber. This means that as a contrast to the eucalyptus, the banksia is actually increasing in size of population. The abundance of this plant however, is only one of the factors when deciding if it is a viable diet source for an animal. Nutrient content must also be addressed. The banksia pollen is extremely high in protein (42% according to a journal published by the National Society Oikos). Protein is a source of long-term energy that could ultimately cause the koala to be more active and be able to find more food. Other marsupials eat this plant as their main food source: the Eastern Pygmy Possum and the Honey Possum. These animals are proof that this plant is a sustainable food source that can support a population of animals.

Wattles: Wattles are the most abundant plants in Australia and can be found in many other places with similar climates such as Brazil and parts of Africa. The aboriginal people to make flour for consumption meaning that they must have nutritional value use the seeds of these plants. Sugar gliders and many types of possum eat this plant and the many bees and insects that hang out around the plant. This diet is suitable for other marsupials, but further tests would have to be done to see if it is suitable for the koala.

Grevillea: Known for its sweet nectar that was used by the aboriginal people, the Grevillea would be a fine source of carbohydrates and sugars for the koala. However, some species of Grevillea are highly toxic, containing cyanide. This would pose a little problem for the koala, but knowing how intricate their digestive system is, they could probably digest the poison. Eucalyptus already contains comparable amounts of toxins, but further tests would have to be done to find out. The presence of cyanide however, can be a benefit because it limits competition for the food source. Not many animals are daring enough to eat a plant containing bright red flowers and cyanide.

Melaleuca: A plant used in many tea tree oils and pharmaceutical products because of its healing ability, Melaleuca would be a beneficial plant to eat because of these health benefits for the koala. This plant is very available in Australia and is also invasive in Florida as well. There isn't much competition for the plant because it isn't eaten by any other marsupial or mammal.

All of these plants would be heavily available and nutritious sources of food for the koala because all can be found in Southeastern Australia and most of them produce food similar to the eucalyptus without the downside of toxins (with the exception of Grevillea of course). The benefits of this food switch could produce very promising results. The species, like the polar bear, would be able to survive on new types of food and adapt to the climate change. They would still eat eucalyptus as well, but these food sources could complement it, or even become more prevalent in the koala diet like caribou and geese were for the polar bear. This could radically change the overall behaviour of the koala as well. With added protein from the Banksias, the koala will be able to have more energy in the long term, which will lead to being able to avoid starvation and losing their habitats to urbanization. Koalas would be able to move away from deforested areas instead of staying there hopelessly. This would lead to an evolution in the species of koala that we are already seeing in the polar bear and many other species that are adapting to climate change with changing diets. By eating a more prevalent and drought resistant plant such as Wattles would help koalas withstand long periods of drought that are common to the Southeastern Australia area. These changes could ultimately increase the lifespan of wild koalas, which is currently around 10 years. Koalas in zoos live for around 20 years, double that of wild koalas. What is happening in zoos could be similar to what will happen if koalas adapt their diets. In zoos, a koala's teeth are preserved and cared for because eating eucalyptus and extracting fibers wears them down. Without the need to heavily extract nutrients from plants like the Banksias, koala teeth won't decay as fast and they will be able to avoid the number one cause of death, which is starvation.

Zoos can sometimes provide a way to view nature that is helpful to the ultimate survival of it. Zoo animals usually live longer than wild animals and can be closely monitored as to learn more about the behavior of the animals. Zoos are safe places to test new dietary items such as the plants listed above to implement into the koala's diet. They could be implemented in small amounts as to test if there is any other viable food source they could introduce to the koala. Such controlled environments have proven themselves in the past with introducing new animal behaviors that can be implemented in the wild. This was the case for testing the teeth of Rhinos to find which new sources of food they could introduce into their diet in a South African zoo.

But as any new learned behavior, it takes a long time. The phrase "you can't teach an old dog new tricks" rings very true in the animal kingdom. Koalas are one of the most ancient species left in Australia and to completely change their diet would be to an extent, reversing their evolution. To teach the species of Koalas to eat a new diet would require the spreading of learned behavior from parents who were taught it to children of the next generation. This would involve zoos teaching a small group of fertile young koalas how to eat different foods because animals learn more when they are younger. Those koalas must then be released into a controlled environment with the new food present in order to teach other koalas this strategy of eating. This plan would be like training koalas in boot camps how to survive the wild, which is pretty much what conservation centers and zoos are already doing. The trained koalas must then be released into the wild so they can share their new learned behavior to the entire population of koalas. But like anything that is a natural process, this would take time.

A study was done by Bournemouth University concerning this field of interest. For this study, the researchers produced a systems model of time budgets revolving around core activities of animals including periods spent feeding, moving around, resting and engaging in social interactions. The team then considered abiotic variables on these activities such as temperature, seasonality and rainfall and biotic variables including the composition of an animal's diet, its body mass, the quality and distribution of vegetation in the habitat and forest cover. Other factors included the size of any animal community groupings and susceptibility to predators. This study showed that it was a possibility for certain species to adapt to climate change by changing locations or diet, but that it took a very long time. The researchers claimed that animals didn't want to upset their normal behavior and didn't have time to go out of their way to eat new foods.

Animal behavior is always a point of contestation between researchers as they are finding more and more about how unique an animal's mind is, or how similar it is to a human. Nobody yet has been able to fully understand the animal mind, just like nobody has been able to understand the human mind. With over a 100 billion neurons firing in your brain, how could we possibly ever be sure of anything concerning the brain. Humans have however, already mapped out the human genome and are now working on the epigenome, so I wouldn't put it past them to map neurons in the future. Even so, we still have much to learn about the animal's mind and behavior because it is such a new field of research. I hope to someday contribute to this growing field of study, but currently am stuck in a classroom just learning about it.

A new diet for the koala could vastly help them survive the growing threat of global warming, but would also help them to change as a species to become more active and more of a keystone for the

Australian ecosystem. Koalas are already one of the largest sources of ecotourism in Australia, so if they could also contribute to the ecosystem physically it would boost Australia's beauty even more. Koalas have never been trained to do anything like a dog, so if this plan is implemented it would be a first of its kind. The results may not be promising the first few years of the trial, but training takes time: immediate results aren't expected. Global warming, although threatening now, will not destroy the planet anytime soon, so we have time to wait. If this option however, does fail, there is still a chance that koalas will learn this new behavior on their own. Animals and humans alike have a special ability to sense danger in the environment and adapt to that danger with change. Like the polar bears eating caribou because of decline in seals and like the koalas drinking out of fountains because leaves contain less water, the koala can sense the danger of lack of food and adapt. Nature has a weird way of restoring itself and advancing the life of certain species it deems important. Without human intervention, the koala may fare well.

But we unfortunately don't have much time to let nature work. With eucalyptus on the decline in water content by 80% and deforestation and urbanization taking place, koalas could soon become endangered for reasons that could've been prevented. The leading cause of death for the koala is starvation: this must be stopped.

Chapter Three:
Behave like a Koala

The behavior of a koala is definitely unique amongst all else. They sleep for 16- 22 hours a day, only eat a certain type of food, are solitary, meaning they don't move much, and have pouches for their young. Koalas also prefer to be alone rather than in groups, a strange behavior for animals. All these behaviors are indeed unique, but many people still do not know the reasons behind the different behaviors koalas exhibit. Such questions will be answered in this chapter.

Sleep

The most distinct feature of the koala is that it sleeps for so very long. The longest sleep time monitored by a koala was 22 hours a day. That means the koala was only awake for 2 hours! In that time, the koala must eat, care for any young and eat some more. This daily habit seems very redundant and a waste of time for the koala, but it is actually necessary to live. Most other herbivores such as antelope don't sleep at all, or at the most 5 minutes. Carnivores are the opposite. Lions sleep for an average about 20 hours a day, similar to the koala. This is because they expend a lot of energy hunting and digesting their large prey, so they cannot waste any energy simply roaming around. Sleep is nature's best method of conserving energy because when sleeping, the body is barely more lively than dead. When you sleep, your muscles are relaxed, heart rate drops, breathing is slowed, and energy is restored. The only thing that is active fully is the brain, racing with dreams and ideas when you sleep.

The same is true for animals as well. A study done by Hugo Spiers at the University College London suggests that animals can in fact dream. Spiers and colleagues have found that when lab rats are shown food and then go to sleep, certain cells in their brains seemed to map out how to get to the food. Another coinciding study done by Patrick McNamara also proved the fact that animals could dream. He altered a cat's brain to remove the mechanism that inhibits movement during REM (rapid eye movement) during sleep, that is the phase during the sleep cycle where dreams happen. McNamara found that cats got up and sleepwalked around the room, appearing to be interacting with invisible objects. They had appeared to be stalking prey during their sleep, implying that they were dreaming (probably about being a lion or something). Rats were also seen dreaming about a maze after running through it a few times through. Their brains were monitored during sleep and compared to the brain as they were running the maze. They found that both were identical: they could even figure out which part of the maze the rat was on.

Who knows of all the things koalas could be dreaming about during their twenty-hour sleep period? Possibly dreaming about eucalyptus, one of the only things they come into contact with in their day. The reason that koalas sleep so much is because of the nutrient content of eucalyptus: next to nothing. The koala needs to conserve energy, so they slow down their metabolism and they do this by sleeping. Less energy will be consumed by cellular respiration, the metabolic pathway of turning glucose into energy. Cellular respiration converts the glucose from the food source into two pyruvate in the first step of the process: glycolysis. This occurs by the oxidation of glucose, stripping it of its electrons in order to break it down. After pyruvate is made, it must be slightly altered in order to enter the citric acid cycle. There it is stripped of its electrons by the molecules NADP+ and FAD+. These polar molecules have an affinity for electrons and will transport them to the electron transport chain. The electron transport chain uses oxygen in order to drag the electrons along the chain and push the gradient of hydrogen ions against the gradient. This powers the ATP synthase pump, which makes 36 ATP: the energy that all cells use. This process can be sped up by heat, or slowed down by coldness. This is why koalas sleep: to lower body temperatures and muscle movements so they can eat less and still live.

You may think that global warming would increase this process because the world is becoming hotter, but this is not the case because koalas and most mammals are endotherms, meaning they maintain their own internal body temperature. This is done by sweating when they are hot and shivering when they are cold. Thermoregulation is all a part of a bigger idea of maintaining balance in one's body called homeostasis. Homeostasis is the basis for life, as most organisms will die without it.

Solidarity

The cliché picture of a koala will probably have them hugging a tree (trees provide coolness). Usually, koalas will stay in the same tree their whole life. They rarely express social interest in each other unless they have a baby, or are trying to find a mate. Many other pouched mammals such as the kangaroos are very social animals and will often travel in groups. Many herbivores are also social animals. But as we have seen already, koalas are the one exception to all the natural rules of herbivores. They have already broken the rule that most herbivores are fairly active during the day and barely sleep at all: what is one more rule to add to the list.

The main reason that an animal would become solitary is to limit competition between other members of the same species. Common solitary animals are jaguars, bears and tigers. Most predatory animals are solitary because there is a limited food supply high on the trophic level of an ecosystem. For koalas, this is not the case. There is no competition among koalas because they sleep too much to have that problem. There also currently isn't that much of a food shortage to compete against, so the reason for their solidarity is pretty much nothing. Based on these reasons, koalas must be one of the most antisocial animals of all time. They are choosing to be solitary because they dislike the presence of other koalas, not because of food competition.

Solidarity amongst animals is actually being used currently to teach us more about autism because both people with autism and solitary animals have lower levels of connection with other species. Animal models such as this are usually used to learn more about human diseases because they are so similar. Both types have a decreased level of separation stress and body expressiveness. Scientists have found that the main reason for this solidarity is lower levels of oxytocin in the blood plasma. Oxytocin is released during social interactions and is the hormone that keeps us wanting more interactions with other people. It also determines how we respond to social problems or stimuli. A major contributor of solidarity and oxytocin levels is the early development environment. Where a baby animal or human grows up will greatly affect the levels of oxytocin in the blood. If babies separate from mothers at birth and are left to fight on their own, they will have lower oxytocin levels and be less sociable. The more that a baby is nurtured and cared for, the higher its oxytocin content.

For koalas, very unsociable animals, oxytocin levels are very low because of their chemistry composition. Their young are only nurtured for about a year, which could contribute to the solitary behavior, but it isn't quite as short as a shark's nurturing period, which is close to none. This means that the koala must have evolved to be less sociable because it was advantageous at one period in time. Probably a long time ago when there were many predators and it was beneficial to stay put and hide, rather than move about and converse with other koalas. Solidarity was an evolved trait for the koala, one that was beneficial in the past, but may be killing them now.

Solidarity, as well as sleeping for too long, poses many problems for a population of koalas. If their home happens to be deforested by the ever-growing eucalyptus industry, they will probably die in the tree that they have been in their whole life. They are very much like the musicians that went down playing on the titanic while it was sinking: they don't want to move.

Vocalization

Many can recognize the songs of birds, or the purr of cats, or bark of dogs, but many don't know that koalas also make distinct noises. A koala's call sounds like a loud grunt, or a squeal for the young ones who haven't developed the vocal chords yet. This call is usually used to assert territorial dominance or to protect their young from other koalas trying to get into the area.

In an interview with the professor of koala ecology at the University of Queensland, Bill Ellis, the koala code was finally unleashed and koala communication was brought to a forefront this year. He put tracking collars on female koalas and noticed that when a male bellows, a female usually responds meaning it has something to do with mating. Mothers also responded to babies making a "yip" sound when they are lost. In distress, the koala call gets much higher pitched and sounds similar to your friend after you step on his toe with your steel toed boots.

Many different types of dolphin and whale recognize each other by their calls. This is true for songbirds as well, as each bird has a unique song that they will sing to attract females. This is not the case however for koalas. Koalas, as previously noted, are solitary animals and do not hang around other koalas outside of their family very often. Calls are not used to recognize individuals because they are not out looking for mates or other koalas to hook up with because it is against their nature. Many times, koalas use bellows to tell other male koalas to get out of their area. But often, the koala giving the calls isn't willing to back up his talk. Koalas aren't very aggressive animals and it is very rare to see them attack anything, so they are pretty much just taunting other males. Koalas are like the bully who only uses verbal insults without backing up his claims with physical strength (if anyone wants to send me a picture of a koala fight, go right ahead).

In a koala society, because of their pacifism as a species, there is no dominant "alpha male" position. Female koalas breed with a different male each year and no single male is dominant when it comes to choosing. This is one of the things that makes koalas unique in the animal kingdom because most animals have some sort of social structure, at least most mammals do.

Recent observations on animal vocalization suggest that some animals have actually learned their calls through perception of other animal sounds and many years of evolution. We all know that parrots can repeat sounds that they hear from humans, and dolphins, elephants and bats can imitate sounds they hear as well. It is a phenomenon that happens when an animal hears and perceives a certain sound and is then able to trigger a neuron in its brain to remember and repeat the sound. This complex process furthers the hypothesis that animals are not only learning from human beings, but that they in their own selves are intelligent. After reading many different books on animal intelligence such as "Beyond Words" by Carl Safina and "Are We Smart Enough to Know How Smart Animals Are?" by Frans De Waal, I am convinced that animals are intelligent beings of nature and some are even smarter than us. Animals capable of making tools and understanding sound can do many of the same things we do in our everyday lives.

Scent Marking

Many have seen dogs mark their territory through scent marking, a self-explanatory phrase. Wolves are actually one of the most prominent animals that perform this behavior because they are extremely territorial animals. They can smell the scents from miles away and can therefore, avoid fights with neighboring packs. Scent marking is key to maintaining peace and order within animal societies, just as land borders, border patrol, laws and regulations, and constitutions are in maintaining ours. Without borders, there would be no order to the world, as every country would just assimilate into one. Fights over land would ensue because there is no firm line between them and it would be much harder to control a territory without borders than one with them. This is very true for the animal kingdom as well.

Koalas are one of the animals that participate in this phenomenon and they do so in order to maintain borders and mark territory. Koalas aren't big fighters, so this is one of the ways that male koalas can assert dominance in a family by marking territory. Males who place more markers have also been noted to produce more offspring according to Dustin Penn from the University of veterinary medicine in Vienna. He states that this is because scent marking takes energy and attracts predators, and low-quality males don't have the energy to perform the task. Penn also discovered that the more intrusion into a territory the animal receives, the more it will mark its territory. This means that koalas living in densely populated areas such as Queensland will produce more scent markers in order to claim their territory and assert dominance. Scent markers however, could be interspecies, meaning that they would also be in effect for dingoes and other Australian animals. Further research must be done however, to prove this hypothesis.

One rare use of scent marking in koalas is by females during mating season. Females will use scent marking to attract males, who will then enter the territory to meet the females.

The Pouch

Little is actually known about the pouch that wields a koala's young except that it keeps the baby protected and safe. I can infer however, that based on epigenetic research done on rats, the pouch plays a much larger role in the development of baby koalas than just protection and safety. The University of Utah studied rats in a contained environment. One group licked their young; the other didn't and completely ignored the baby rats. The results helped form our concept of modern epigenetics, which means above the genome. Epigenetics is the study of environmental factors that can affect gene expression by methylation. Methylation occurs when fats called Methyl tags bind to chromosomes and unwind them so they can be expressed through transcription and translation.

When the chromosome is closed, there is no way that it can be split by the enzyme helicase in order to be used as a sample for replication or expression. In the rat experiment, they found that rat babies who were licked a lot, were less stressed and didn't have anxiety. Rats who weren't licked and were ignored by their parents, were the complete opposite: anxious and stressed. At first, there was confusion because epigenetics was a new science at the time, but it then became clearer. When a rat is licked as a youngling, its brain releases chemical signals, hormones, in response to the licking sensation. These hormones, if enough of them, can bind to Methyl tags, causing them to fall off. This means that the gene for anxiety will not be expressed because it doesn't have Methyl tags on it. Rats who are not licked, will not release the chemical and therefore, the gene for anxiety will be expressed. This can actually be a good thing if a rat lives in a sketchy environment with a lot of potential dangers such as hawks.

Koala pouches may serve the same purpose as licking did for the rat. Since koalas live in an environment that isn't very dangerous and they have no natural predators, they don't need to be anxious and waste their energy being alert all of the time. Their pouches may serve as an epigenetic way to lower the levels of stress that the koala experiences and to silence the very same gene that the rats did when they were licked. An experiment could be done where a koala is left out of its pouch to develop on its own, and the other develops normally in the pouch. This could see if the one without a pouch saw a major difference in behavior and we could see if it is really an epigenetic phenomenon taking place.

Koala behavior is truly a unique mystery that is yet to be solved completely. There are many questions to be asked, experiments to be done and discoveries to be made. I hope that my suggestions will be of some use to the advancement of the knowledge of the koala and its mysterious behavior. Koalas will always be a unique animal among the animal kingdom and will always be an exception to the natural rules we have seemed to make govern an unnaturally complex world.

Chapter Four:
They're All The Same

When you think about animals such as domestic dogs, caterpillars and mice, you can't help but notice how much diversity there is between the species. One mouse that is located in a desert may have a brown coat, while one in the forest may have a darker black coat. Two Dalmatian dogs won't have the same amount of spots in the same locations. You'd think this would be a universal truth among animals because we are so used to individualism and everything looking slightly different, but this is not the case.

Inbreeding is a very common practice in the animal kingdom and is the practice of animals breeding with their relatives. This is usually caused when animal populations are scarce, or it can be caused by a learned behavior. For koalas, inbreeding is the most common type of breeding and it is causing the species to have a very low genetic diversity.

Scientists believe that this inbreeding behavior was one learned by koalas because of low populations in the 1800s. Aboriginals hunted them for their fur, and Europeans did the same when they came to Australia. These events caused the number of koalas in the wild to diminish by a lot, thus making inbreeding one of the only options. Even though koalas today are more numerous and inbreeding isn't a necessary means for survival, they still perform the behavior pattern. This is probably because the behavior was learned and passed down from generations as a form of behavioral evolution.

Scientists tested the diversity among koala genomes by comparing them to those in museums. The ones in museums were from different regions and different time periods than the ones tested against. These totally different regional species however, did not differ from the modern day koala at all. They were only about 3 letters in their genetic code apart.

If you go to Australia and see the koalas there, you will probably notice that they all look the same, unlike mice or other species with high variation. This may not seem like a problem, but the results of low genetic diversity are very bad for the species. Low genetic diversity causes a species to have a very hard time adapting to new environments like that caused by climate change or global warming. This is why koalas are starving themselves and getting insufficient nutrients from eucalyptus, despite the hot weather and dying food supply. They have been unable to adapt to this new environment because of this low diversity genetically. Low genetic diversity also allows diseases such as chlamydia for the koala, to spread rapidly and for the species to be unable to fight it. This is because these diseases target the same genes on all of the organisms. Less diversity means the same disease can target all koalas. If a genetic virus was to target a certain koala, it is most likely that the whole species would become wiped out.

Low genetic diversity has been the downfall of many different crops over a historical time period as well. The Irish potato famine in 1845 was partly caused by a disease that infected the potato plants and since they were not genetically different, they all died. The effects of having low diversity could've also been the cause of the extinction of so many species of dinosaur. Mammals have been noted to have higher diversity than reptiles, so a genetic disease that didn't wipe out mammals because they were so diverse might have wiped out dinosaurs.

Genetic diversity is what causes life to thrive, as you will see later in the book, but the lack of it could be the downfall of a species. An example of this is the Tasmanian devil, a marsupial that lives on Tasmania and looks like a big black rat. This species has been noted to have very low genetic diversity and because of it, devils everywhere are breaking out in facial tumors that are caused by a disease. This is causing them to be unable to eat or swallow and causing many Tasmanian devils, a keystone species, to die. Now they are considered endangered species because of these nasty tumors. The tumors usually kill the devil before it can breed, which means the species is declining very fast. Tasmanian devils are now being taken into sanctuaries and reserves where scientists hope they can recover and breed. The future of this species relies on the success of these programs.

Koalas could fare the same fate as the Tasmanian devil because of their low diversity, or fail to adapt to climate change. It seems as though a time is coming when all species on earth must adapt or die, and I plead that the biodiversity on earth does not perish because of it.

Chapter Five:
Koala Cousins and Fake Trees

We have already explored in depth what makes a koala unique and how that unique behavior causes many problems for the Australian animal. What we still haven't explored however is how the koala got to be such a unique freak of nature, one that abides by no known rules we have set in place. So the question for this section would have to be: how did the koala get here, and why?

Koalas are thought to be descendants of another koala-like species called Nimiokoala greystanesi. This species had a little bit of a longer snout than the normal koala today, but looked very similar. It also lived in the same area as the modern koala and had similar diets of leaves. This ancestor lived a long time ago proving that koalas frankly haven't changed much over the years. This either means that they are perfectly suited for their environment (which we can clearly see they are nowhere near perfect) or the idea of evolution and ever changing natural selection doesn't apply to koalas, or any life to that matter. In this chapter we will be explaining how koalas should've evolved, but unfortunately did not. To do this we must look no further than the closest relative to the koala family: the wombat.

The Physical Difference

From a far enough distance, one may mistake a wombat for a koala. They both are small animals that are about the same size and shape. Both can be found in Australia and they both love eating plants as well. There is almost no physical difference between them.

Koalas do have slightly bigger noses and ears to make up for their other senses that they rarely use such as taste and movement (there isn't much point in taste if you only eat one thing your whole lives). Other than that, you could definitely tell that these two animals were related. From the outside they are similar, but if you were to match them up in a cage and study them for a few weeks... you'd see where the real difference kicks in.

As we know from the previous chapter, koalas love to climb trees and will sometimes stay in the same tree their whole life. Wombats on the other hand, do not climb trees and stay grounded. Koalas are terrible swimmers (if you need proof, look up videos of people saving drowning koalas on YouTube), while wombats are fantastic swimmers and even have web-like feet to propel them in the water. Koalas are some of the sleepiest animals and sleep for up to 22 hours of the day. During this time they are not alert and moving fast, and even when they are awake, they are very slow movers. Wombats are the exact opposite and are very agile and alert mammals. The do move quite awkwardly at times, but for the most part, they have great agility when they need it to escape predators like the dingo. Wombats can even run up to 25 miles per hour, which is faster than the fastest Olympic sprinter.

One thing that both animals do have in common though, is that they are both solitary. Wombats burrow their homes in the ground and will usually not socialize with other wombats just like koalas don't socialize. This must be a beneficial trait for Australian mammals because not many of them actually do socialize. As opposed to monkeys and apes, this is an interesting point of difference. Australia is one of the most dangerous places in the world and is home to some of the most dangerous and poisonous creatures alive such as many great white sharks and is also home to the 30 most venomous snakes in existence like the eastern brown snake and tiger snake. This dangerous environment is like a constant battlefield. Unlike the peaceful tranquility of savannah habitats or desserts, animals living in Australia must fight each other in order to gain the advantage of territory and food. This is probably the reason for solidarity amongst Australian animals. The environment you grow up in greatly affects your epigenetics as seen in the rat experiment from the previous chapter. Studies have also been done to prove that if a parent is addicted to drugs, the child will most likely have an addiction as well. This effect may hold true for the animal kingdom as well. If a parent is constantly on the run from predators, the baby will have the same anxiousness and high levels of stress. Solitary social behavior is just a side effect of this reaction. If you are scared and alerted by everything, then nobody is your friend. Pretty much, in Australia it is every animal for itself: trust no one.

Now looking back at wombats... they use scent markers as well, but for a different purpose. As koalas use them for mating and territorial borders, wombats use them to mark hunting grounds and to warn other wombats not to eat from their finds.

One thing however, that wombats do very similar to koalas, is conserve energy. Wombats build burrows in the ground and usually sit in them for a majority of the day. This is very similar to a koala resting in a tree and serves the same purpose of slowing down the heart rate and respiration rate in order that they can survive with less food. They do not need to do this however, because food is abundant, but choose to. This choice to conserve energy is very fascinating because it shows that animals have the intelligence to save for the future just as humans do. This could also point towards evidence supporting that animals have "theory of mind", meaning they are aware of other animal's needs and desires. They may be leaving the food so they could save it for other wombats or animals to use, but one thing is for sure: this behavior has caused the carrying capacity for wombats as a species to increase because they require less food to operate. This is a difference between wombat and koala that attributes an increased level of thinking to the wombat that the koala frankly doesn't have.

One problem that wombats do face however, is heat. Wombats have no sweat glands and have a lot of trouble regulating their internal body temperature. Like most mammals, wombats are endotherms, meaning they maintain their own body temperature in response to the environment by either sweating or shivering. Since the wombat has no sweat glands, it can only do about half of the thermo regulating by itself. Wombats have been reported to sit inside of water holes in order to cool off on a hot day, but most of the time they are able to stay cool by just staying in their burrows. At night, when the wombat must get food, it forages in the dark for up to 3 hours (talk about a long meal). People have reported wombats walking up to 1.5 miles in order to find food and bring it back to their burrow to eat

Koalas are known to stay in their same tree from birth, but this is not the case for wombats and their burrows. A wombat territory can have as many as 14 different burrows in it that all the wombats in the territory share. A wombat will visit 3 to 4 different ones in a night, which shows how much more active it is than a koala.

Although appearing similar from a distance, koalas and wombats live completely different lifestyles. Each has a unique behavior that is centered on the principle of every man for itself, or in this case: every marsupial for itself. Next we will explore what neurological differences cause these behaviors and what may be the sole cause of why a koala isn't as active as a wombat.

Koala Mind Games

The brain is one of the most vital organs in the body and is what makes most species diverse and unique. Each brain is geared differently with its own purpose. But it is undisputable that some animals are much smarter than others... and this is definitely not the case for the koala. Koalas have been recorded to be one of the least intelligent animals in the whole animal kingdom, and it is mostly caused by how their brain is wired.

Koalas unfortunately aren't that bright of animals. A population of them in Queensland almost killed themselves because they had eaten all the food in the area and didn't know where to go and get more. This sounds like something a 4-year-old human would go through when they lose their mom at the grocery store. This makes the koala's brain very childlike (without the playing aspect) and adult koalas will usually maintain quite a low level of intelligence their whole lives. This is not to discredit the koala in any way because koalas don't need to be very smart to survive. They only need to know what food looks like and how to climb to maintain their diet and habitat their whole life. Because smartness isn't needed in a koala's life, it isn't smart, which poses a big problem for the species. While not being intelligent was beneficial in the past for koalas, now there are much more complex dangers that koalas face such as food shortage, deforestation and habitat loss. An experiment was done on koalas to see if, in the case of food shortage, they would eat eucalyptus off a plate, and none of the koalas even knew that it was the same food from the tree. This is problematic because the koalas will have a lot of trouble staying alive and adapting to new habitats and food sources if their homes are destroyed, which is why it is extremely necessary to push for the protection of koala homes.

A koala not being intelligent isn't all due to their brain composition however. Remember the wombat, its closest relative? I thought about the potential change in brain chemistry between organisms, so I decided to run some tests myself using a fantastic genetics program called BLAST. This program sequences genes and their nucleic acid bases in order to see how similar the gene is between different species. It does this by looking for base pair differences such as adenine to thymine and guanine to cytosine if it is DNA. Once this is finished, it brings up a subjective number or percent of similarity between the two genes. I performed this test on the BDNF gene of the koala and the wombat, which is one of the genes that compose the brain. I found that the results proved very contradictory to the fact that koalas aren't very intelligent and wombats are smart enough to remember every hole that they have dug and been in. The gene was 94% similar between both species, with only a few differences in base pairing. This means that their brain composition is almost identical. This result only raised more questions and led to more research, which I will soon address.

Food for Thought

Based on the BLAST research done by me, we can conclude that the difference in behavior between a koala and wombat isn't caused by genetics, or is it. We have already discussed epigenetics and how a koala's environment has caused it to behave like it is, and how a koala pouch may serve to make a koala more docile by releasing the methyl tags on the baby. Epigenetics however, can apply to more than just how the environment affects your genes: food can also affect the epigenetics of an organism.

Research done at Lund University has shown that type two diabetes is in fact, an epigenetic disorder and it is based on whether your parents had the disease. The scientists at the university discovered that there were epigenetic changes in 800 different genes that they studied. They also noted that these epigenetic changes could be passed on throughout generations. This is the same with smoking, as children who have a parent that smokes, especially a mother, are more likely to become addicted to smoking because they have their epigenetic tags. Also, the university noted that when a woman is pregnant, what she eats could affect the baby's epigenetics as well. Epigenetics can be ever changing, even to the point of being inside the womb. But they can even change beyond the womb, which is what is happening with koalas.

Koalas are known for eating eucalyptus from gum trees exclusively. Eucalyptus is a very interesting plant that has been used numerous times for clinical effects such as reducing inflammatory response, reducing pain and swelling and lowering blood pressure. This may seem similar to another hormone that is commonly associated with stress, which is cortisol. Cortisol does many of the same things as eucalyptus, but is a hormone rather than a plant. In the scientific journal, cortisol has been researched to affect the brain's function and plasticity. It can permanently alter brain chemistry and suppress certain neurons from firing.

This same effect could be happening to koalas that eat eucalyptus, a plant similar to cortisol. They could be permanently altering their brain function and plasticity by eating eucalyptus. This food source could be epigenetically changing them just like epigenetics changed a person's likelihood of receiving diabetes or an addiction to smoking. This, combined with the fact that mothers eat eucalyptus during childbirth is probably the cause of the uniqueness found in koalas that isn't found in wombats. This may be also why koalas only eat eucalyptus and nothing else: they think it taste better, or addicted to it because of the epigenetics of their parents that have been passed down. It is just like how humans pass on certain tasting genes like TASTR, that koalas pass down their affinity for eucalyptus. Based on these observations, research concerning the diet of the koala would have to be done on an epigenetic scale, which we have not fully begun to grasp yet. But nonetheless, there is a human epigenome project in the works right now, trying to break the code beyond the code. Koala behavior could give insight into this research and benefit from the research to learn about the true diversity of species and what makes them different and unique.

The only physical and genetic difference between wombat and koala is diet. There are too many differences between the behaviors of the two to not take into account that their diet has changed them, shaped them into their own unique beings. Causing an astounding epigenetic difference.

Could Fake Trees Work

So where does all this research come to an end? Koalas are in need of a new food source, or one that is plentiful at the very least. Koalas, as we have learned have an addiction, or strong desire for eucalyptus that cannot be simply changed by introducing a new source of food. But what if we could introduce a version of the eucalyptus tree that is both healthier and contains more energy for the koala, and it more plentiful that the eucalyptus is right now. The good thing is, this kind of eucalyptus already exists... the bad thing: it is an invasive species.

Genetically engineered eucalyptus was accidentally introduced and legalized in the southern states of the USA. It is currently being debated on whether it should be able to be planted here or just cut down because it is highly damaging to the forests. The genetically engineered eucalyptus is taking all of the water out of the soil and also is highly flammable. This causes the other trees in the forest to receive less nutrients and have a higher risk of burning down. These trees grow 3 times faster than normal eucalyptus trees and produce more seeds and nutrients.

The invasiveness of this tree is a problem in the United States, but in a place where eucalyptus is desperately needed, it wouldn't be such a bad thing.

The flammability of the eucalyptus could be controlled because of the koalas keeping the leaves low by eating them. The extra energy and seeds produced would also help the koala keep the forests of Australia populated despite the growing use of eucalyptus for wood and such.

This all sounds nice, but the question is: could this actually work? Would koalas actually recognize this as an edible plant that tastes the same as normal eucalyptus? The answer is probably not. Koalas aren't insanely smart, but they are smart enough to distinguish between eucalyptus on their favorite tree, and other plants that look similar. Koalas have been reported to not even eat eucalyptus if it is given to them on a plate because they don't recognize it. My initial guess would be to hypothesize that the koala wouldn't eat this new type of eucalyptus, but if they did, it could really change the trajectory of the species. I think it is definitely something worth spending a little time researching as a potential savior to the food problem.

Koalas will however, remain to be unique in the animal kingdom. But their food may not be the only problem that is threatening their species. In the following chapters, I will explore another threat to Australian biodiversity and quality of food that is out of the animal's control. It is a problem so vast that is literally covers the whole entire continent of Australia... and the whole world as well.

Chapter Six:
The Law-Breaking Koala

After becoming thoroughly familiar with the koala and all of its attributes, we will now revisit the first article that got me writing this book in the beginning. That article was one published by National Geographic, and it was explaining how some Koalas have been sighted drinking water from a fountain due to global warming. This was unusual because Koalas don't drink water and get it all from eucalyptus leaves. The name koala literally means "no drink". But they had to drink at one time right? They weren't always getting water from leaves. In this chapter we will explore the mystery of the koala drinking behavior and a possibility that evolution may be working backwards.

One previous ancestor of the koala about 15 million years ago was *Litokoala kutjamarpensis*. This koala ancestor was much smaller, about ⅓ the size of a koala, and ate a nutrient rich diet that included water. This diet enabled it to be more active than the present day koala, which is also proof that the diet is the key to a koala's sleepiness. By means of a long period of time and some natural selection, this koala ancestor became extinct, or adapted to be larger like the modern day koala. Between now and 15 million years ago, the koala has lost the behavior pattern of drinking water, or so we thought.

The modern koala must have retained the ability to identify water as something that they can satisfy their thirst with, or else the Queensland koalas wouldn't have known to drink out of the fountain. This ability was preserved for millions of years and is now vital to the survival of koalas in the hot warming climate. This resurfacing of the ability to drink water however raises a few questions. Was this an ability they always knew, just never used? In this case, LaMarck's theory of evolution based on use would be broken. Or was this ability just discovered by the koala? A question that if answered yes, could change how we view evolution and how it can work.

Recent studies show that evolution cannot go backward. A group of scientists from the University of Oregon tried to "un-evolve" a modern protein into a more ancestral one by editing the genes. Their research found however, that genetic blockades had arisen and had prevented the protein from going back to its previous stages. The team did further research on another protein that was a receptor for glucocorticoid. They created an ancestral version and normal version and compared the genetic sequence. They found a few key mutations that were different and caused the normal one to produce with an updated hormone. The scientists then tried to recreate the original ancestor protein by changing these mutations back, but they wound up with a dysfunctional protein that was pretty much dead.

The scientists suggest that this is because as the ancient protein evolved, five other mutations made subtle changes in the protein's structure that were incompatible with the primordial form. It is the same effect that you have when you redecorate your bedroom. If you move your bed and put your desk there, then want to move your bed back to the same spot: you must first move the desk to do so.

This irreversibility of molecular evolution can be applied to larger scaled macroevolution as well, as most molecular processes are applied to bigger ones. Types of control inside of cells such as positive and negative feedback control are the same ones that can be found in ecosystems on a large scale. This finding in microevolution can be generalized for macroevolution as well, but what about behavioral evolution? Behavioral evolution follows different rules than the others because it is very subjective and focused on environmental changes. Behavior could be caused by the outside world, like not going outside because it is raining, or the inside world, not going outside because you are sick. Because there are so many variables that play into this type of evolution, it is very hard to study it and make any strict conclusions. But considering that the behavior of the koala has reverted back to its origins, it could very well be reversing its behavioral evolution.

The second theory that the koala could be breaking is LaMarck's theory of evolution. This stated that if a certain trait was used, it would grow stronger and be passed to the next generation. This seems to explain why dogs have such excellent sniffers, or why kangaroos have such muscular legs, but it doesn't explain why koalas are drinking water despite not having done so in the past. Their trait could have been lost a long time ago due to the lack of use of it according to LaMarck's theory, but it wasn't. This could be a potential point in the rebuttal against this theory, but as with every theory: there will be exceptions. I still do respect LaMarck's theory because of the breadth of animal species it does apply to, but do not think it is a universally applicable theory that can be used as a pseudo-law.

So while we may never know the answer to this acquired new behavior that koalas have gained that was resembled in their ancestors, I think I can come to my own conclusions. Since molecular evolution seems to be irreversible, I think it is safe to say that the trait of drinking water didn't become extinct with the koala. I think that it was a trait that was always present inside the koala, just barely used. This however, doesn't coincide with LaMarck's theory, but could be used as a point against it. I think that understanding that exceptions and counterarguments are available for almost any hypothesis and theory makes this a reasonable hypothesis: that the koala is drinking water because it has always known/had the ability to drink water from its ancient ancestors. Evolution is not reversing itself here.

Chapter Seven:
The Problem Down Under

By now you have grown familiar to the problem of the eucalyptus, which it is losing water content and becoming less abundant in Australia. This is partially due to the lack of rainfall in Eastern Australia and global warming as a whole, and is also partially due to the deforestation of Australian eucalyptus forests and the use of the wood and leaves for ointments and such. But there is a problem with Australian forests that goes deeper that the roots of the eucalyptus trees themselves... the soil.

Soil is where a plant gets the majority of its nutrients such as nitrogen, phosphorus and even water. Without good soil, a plant cannot grow. It is as simple as that. Soils can also have differing pH's that are caused by the environment the soil is in. Acid soil will absorb pollutants from the outside air and also kill plants, while normal soil shouldn't have these effects. The optimal range of pH for plants to grow in soil is 5.5-7.7.

So how does soil maintain these nutrients and pH levels? The answer is through a vast number of chemical reactions caused by bacteria in the soil. Decomposers are small bacteria that live in soils that pretty much break down any type of living matter and recycle it for plants to use. Decomposer bacteria are so numerous in the soil that a teaspoon of soil can contain up to 100 million bacteria on it ranging from 10,000 different species. These bacteria play a crucial role in the planet by recycling resources.

In the nitrogen cycle, bacteria can convert nitrates into free nitrogen that plants and animals can take up as nutrients. These bacteria also have an effect in the carbon cycle by releasing carbon dioxide gas into the atmosphere when they break down organisms in the soil. Some of these bacteria are even used to clean oil spills, as they can break down complex fats and oils.

Is it heat?

These bacteria however, have limits of where they can live. There are three main groups of decomposer bacteria and all live at different soil temperatures. Psychrophiles live at 14-59 degrees Fahrenheit. Mesophiles live at 59-104 degrees Fahrenheit and Thermophiles live at 104-158 degrees Fahrenheit. Heat of soil can be a very bad problem for the growth of an ecosystem because the hotter or colder it gets, the less diverse the bacteria that live inside it are. And the less diverse the bacteria, the slower they will decompose things and the easier it is for them to die off.

On average, soil temperature is about the same as the temperature of the air because it is fairly good at absorbing heat. Where koalas live, on the eastern outback and eucalyptus forests, it is pretty hot, but that isn't the only problem: it is getting hotter because of global warming. The increase of temperatures in the recent years has put the soil temperature at around 80 degrees Fahrenheit. This isn't that much of a problem for the Mesophiles in the soil, but it does kill off most of the Psychrophiles during the hot seasons. This lack of biodiversity in the soil composition could be one of the causes of the eucalyptus not having sufficient nutrients: but in reality, all of these problems simply stem from global warming.

Is it the type of soil?

There are many different types of soil in the world. Some can carry large amounts of water like that in the Amazon Rainforest and some are very hard, sandy and rocky like in the desert areas such as the Sahara. In Eastern Australia, the soil is the latter kind. Classified as Solonetz, it is a very dry soil with a hard clay top layer. It is called Solonetz because that word in Ukrainian means salty soil. Solonetz soil has around 15% exchangeable sodium, which is very high for a soil. Sodium however can be a bad thing for plants and especially the eucalyptus. Sodium has been known to suck the water out of things and dry it out. Think of eating a salty pretzel and how badly you wanted water after it. This sensation could cause the eucalyptus to use more water for itself rather than storing it in the leaves. On top of this however, salty soil can greatly affect the movement of water into the plant through its stem, which is called transpiration.

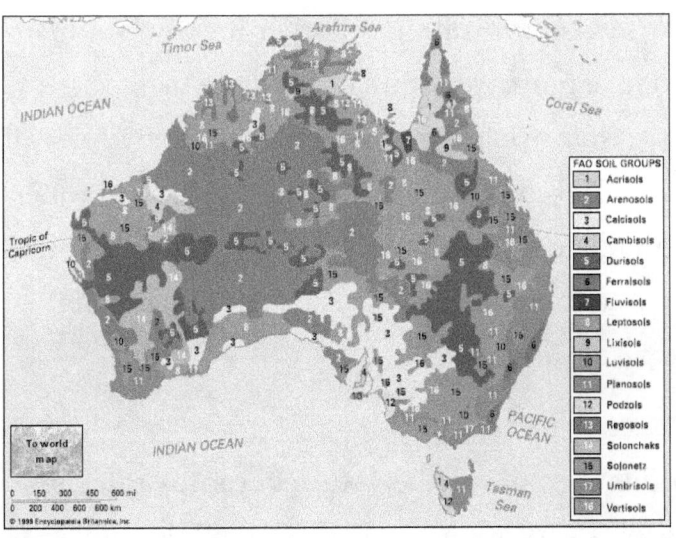

Plants use transpiration, which utilizes pressure to move water upwards through a stem. What the plant does is create an area of lower pressure above the water, so it will naturally want to flow to the area of low pressure. The plant then keeps doing this until the water has made it to all the leaves and the nutrients are distributed. Salt in the soil however, can greatly inhibit this pathway. Salt can cause the water to want to not move towards the area of lower pressure because it would be leaving an area of higher salt concentration. In order to move water, it would take more work on behalf of the plant because they would have to fight the counter force of the salt. This could be the cause of the drop in water content of eucalyptus: it doesn't have the energy to carry water up to its leaves. The consequence of this is less water for the koala.

The other type of soil in southeastern Australia is Planosol, a very abundant clay-like soil that covers approximately 130 million acres of the world. This soil usually houses grasses and small shrubs, meaning it is usually no place to grow trees like the eucalyptus. Because eucalyptus grows on this type of soil, it may be receiving inadequate nutrients to grow fast and to its full potential like the invasive species is in North America.

The Shifting Soil

According to the Australian Bureau of Soil, the soil in Australia long ago was suitable for the diversity of plants and trees, but the soil is now changing. It is absorbing air pollutants and becoming acquainted to a changed atmosphere of heat and dryness.

The soil in Australia had never been used to grow crops until recently because of the shift in type and temperature. This means that this soil is ever changing, and being pushed to its limits. Proof of this "sick soil" as called by the Sustainable Australia Report was found during an audit of the soil quality. The first audit was done in 2000 and showed that the soil health was declining due to erosion, acidification and salinization, or the saltiness of the soil. This only got worse in 2008 when the second audit was completed showing an increase in these problems and required monitoring of the soil. On top of this, Australia is losing much of their fertile soil that could be used for production of food because of inadequate food systems in place and the takeover of urban cities.

Soil health and biology is currently being studied at many institutes because understanding the nitrogen cycle and the role of soil in biology can help make the food industry more efficient and less harmful to the environment.

In this chapter, many different problems were proposed as to what the roots of the eucalyptus problem are. Many of these hypotheses can be tested and many I am in the process of testing myself. If there are any brave scientists among us that want to further explore soil biology I would encourage you to partake in these simple studies. The first deals with heat effects on soil pH and decomposition. We are trying to find out if heat is killing decomposers and possibly denitrifying bacteria as well. Another hypothesis that is yet to be tested is that salt in the soil causes less water to be taken up by the plant. This would be an easy experiment to perform, as you would only need a plant and some salty soil.

Exposing the problem down under the roots of the eucalyptus could be the key to the future of the koala and Australian wildlife as a whole. All animals require plants and grasses to eat, and without soil, those things would not grow or survive. Soil is the fundamental keystone in the Australian ecosystem and every ecosystem at that and shouldn't be taken lightly. Sometimes the dirtiest problems turn out to be the most important ones.

Chapter Eight:
Dingoes on the Decline

Koalas of course, aren't the only animals that inhabit the eastern outback and eucalyptus forests. Many other marsupials are also present such as possums, kangaroos and wombats. Australia is also home to many dangerous insects such as camel spiders and a variety of unique birds such as the Eastern Petrel. One animal that was introduced over 3,000 years ago and has filled a carnivorous niche is the dingo. This dog-like animal was said to have come from eastern Asia by boat.

Dingoes on the outside look almost identical to a common housedog breed, maybe a husky or Shiba. Their story is almost the complete opposite than that of domestic dogs however. They were first domesticated, and then broke free of humanity as a source of reliance. Domestic dogs however, started out as wild wolves and slowly trusted humans and became reliant on them for food and care: thus domestication. This means that dingoes will look like domesticated dogs, but act quite the opposite. Their entire behavior is learned and is based upon a breakaway from the human race, not domestication like a housedog. Rather than comfort and care for humans and perform tasks that they tell them to do, they are wild and aren't a comforting man-lover. As a matter of fact, dingoes actually will attack humans on certain cases and be very hostile towards them.

I have a golden Labrador retriever named Lucy, who lives with us. She is a domestic dog and has been one her whole life. Looking at her behavior change over the years (she is 9 now) I have begun to realize that her personality and behavior is almost entirely shaped by her experience with humans and her interaction with our family. For example, Lucy has learned to hide when we start the sink because occasionally it makes a very loud sound (the trash compactor). Or she has learned that when she rips up her stuffed animals, she gets a new one. Lucy as well as most other misbehaved dogs has also learned to beg for food, because very few people can resist the cuteness of their dog... I am definitely not one of those people. Lucy has also recently developed a huge fear of balloons because I accidentally popped one near her ears. Her behaviors and fears are very much shaped by her experience around humans and are in no way instinctual like that of a dingo. Dingoes are born with fears and behavior patterns that separate it from a normal housedog. Dingoes fear humans not because of something we have done in their lifetime, but because of something done in their history. They pass on these learned behaviors to their young and since they aren't around humans enough to be influenced by them, they remain wild.

So what is dingo behavior?

Think of a dingo like an Australian wolf. They are extremely social animals, unlike the koala, and many exist in packs. Not all dingoes hunt in packs, but the majorities do. Dingoes eat small rabbits, birds and reptiles. Sometimes dingoes have been reported to be attacking larger marsupials such as kangaroos, but it is a very rare occurrence. Dingoes like wolves also only breed once a year, meaning they reproduce very slowly. This could be a problem for the dingo in the future, as we will discuss later on in the chapter.

Many dingoes will also attack herding sheep, which has caused the erection of dingo fences to keep them away from the farms in Australia. Dingoes are also very good communicators and use a variety of whimpers and cries to communicate their situation with others in the pack. Dingoes will occasionally bark in a short manner to communicate distress, which isn't very normal among the dominant species in Australia. Dingoes don't have any natural predators and very little competition for food, so they don't have very many reasons to be in distress.

Dingoes, although not completely reliant on humans for food, are still smart animals and know where to get a quick and easy meal. Ever since their arrival in

Australia, they have been mooching off of human camps for food and scraps. The aborigines that lived in Eastern Australia would often give the dingoes part of their meal and leftovers of their kill. Dingoes are also natural thieves and will steal just about anything that is left unattended. They have been known to take clothes, shoes, food and even people's sleeping mattresses. Dingoes love to play around with humans and will ram into them to try and spark some fun. But proximity to humans has also caused one of the dingo's biggest problems as a species… one that is causing the decline of the dingo, and endangering the future of Australian biodiversity.

The Dingo Problemo

It is rare to see an animal so ancient, stay the same throughout evolutionary pressures. A few examples are the horseshoe crab and various types of sharks, but no mammal has really survived the test of time because the land and earth is ever changing. Humans inhabited the land, not the sea, so it makes sense why sea creatures have been preserved longer. Humans still do affect the sea with acid rain and trash deposits such as the Great Pacific Garbage Patch, but to no extent that they have inhabited and destroyed the land. To give you a scope of the problem, 3 of the 6 million acres of forest in the planet has been cut down. Up to 65% of the Ozone has been depleted by factory carbon emissions and cars. And most of all, over 3,000 animals die every second in the world at the hand of humans.

This wide scale destruction of the world has not been done without some good, but the effects on Australia have been harsh. Overpopulation and crowding through urbanization usually affects the largest animals in an ecosystem the most because they require the most territory to move around and space to live. This is one of the main factors that keep species such as the dingo on the low end of the population spectrum and what makes smaller marsupials such as possums so numerous. By keeping dingo populations lower, other prey of the dingo are free to populate as they please. A dingo being extremely territorial also doesn't help this problem because they will have less territory than is actually present because of other groups claiming land. Most animals in Australia see it as an open lot, but the dingoes see it as many separate nations with distinct borders.

The lack of land however, is a problem faced by all predatory species in today's world, with the exception of sea creatures such as sharks (they face problems of their own). It is a natural environmental pressure that these predatory creatures must face. It is pretty much a requirement of being a predator. This is not a unique issue that is of current importance to a novel writer as myself and therefore isn't worth a chapter of your time. What I am here to introduce however, is a problem very unique to the dingo that is silently wiping the species off the face of the planet. It is a problem caused by humans, but isn't entirely in their control to prevent. For about 10 years now, dingoes have been on the decline due to this one issue: interbreeding.

Scientists say that in 20 years, the pure dingo may be extinct if the trend of interbreeding continues. The International Union for the Conservation of Nature has put the dingo on the vulnerable species list because of interbreeding. Dingoes and domestic dogs that are introduced via house pets by local residents can interbreed with ease.

This interbreeding between species is not uncommon, it is actually one of the ways that species evolve and even classify as a species. Both dingoes and domestic dogs are practically the same species, Canis Lupus, but so is a wolf. This is what allows them to interbreed with each other. But one may ask: Why haven't they done this in the past? The answer to this rests in a simple concept called separation... in particular: Allopatric separation.

The definition of allopatric separation is to be separated by a physical barrier. This is very much the case for dingoes and domestic dogs because before the urbanization and introduction of domestic dogs to Australia, dingoes had no interaction with them. Even when domestic dogs were introduced into Australia, allopatric separation continued to separate the two species because the dogs were mostly in cities where dingoes stayed away from such as Queensland and Brisbane. It was not until recently that this allopatric separation has been broken. It is mostly caused by the expansion into less urbanized areas, where many dingoes are present. An unattended domestic dog in a small-fenced backyard is the seemingly perfect breeding partner for the dingo because like humans, dingoes occasionally like something different and experiencing different kinds of species and stimuli.

So what is a dingo hybrid and why are they so dangerous to the species? Well, since they have been hybridizing with domestic dogs, dingoes have undergone some major changes. Their average weight and size has risen by 20% over the past 20 years, which may start a trend towards an even larger predator that can have the potential to attack various new types of prey. The potential harms of becoming larger will be explained in more detail in the chapters to come, but it has also been proven that larger animals in a species tend to live shorter and develop more muscle and bone problems.

Through breeding, the dingo has also picked up on some of the behavioral patterns of the domestic dog, such as breeding twice a year as opposed to the normal once per year that dingoes have been doing ever since their existence. Because of this, the dingo population has been exploding recently

This may seem very contradictory to the title of the chapter "Dingoes in Decline", but in fact it is not. What I meant by that phrase was the actual entity of the dingo. These hybrids are barely dingoes at all.

The problem lies in the decline of purebred dingoes, who have filled an ecological niche for thousands of years and are now starting to change. According to Dr. Ricky Spencer of the University of Western Sydney "Nowhere on the east coast of Australia, can you find a dingo population that is less than 50% hybrid."

Many scientists also fear that this interbreeding is causing the decline of the pack-like social structure of the dingo. Dingoes, as discussed previously are social animals that hunt and roam in small packs. Domestic dogs, do not exhibit this behavior. They may play with other dogs as companions, but do not constantly rely on other members to hunt for or care for them. They receive all their care from humans. Dingoes are slowly on the path that many wolves took a long time ago to become domesticated into today's housedogs.

The greatness of this problem makes many scientists pessimistic about the issue. Nobody can really control an animal's reproductive habits. No amount of environmental protection can ward off this problem, it might actually make it worse. There are however, certain strides that are trying to be implemented to save purebred dingoes, as they are crucial to the sustenance of Australian biodiversity.

Dingo Safe Havens

Although many are pessimistic about the hybridization of the dingo, there are still many areas that are predominantly purebred. The most prominent of these is Fraser Island. But even "the purest group of dingoes" on Fraser Island is in danger of things such as genetic drift. This phenomenon happens when the population is very small and isolated from the main population. There can be many mutations in the breeding dingoes on the island causing changes and genetic instability in the species. Genetic drift makes it easier for viruses and other genetic disabilities to spread because the breeding population is smaller. They are also contained on the island with no escape from the genetic variants. Genetic drift is what puts the population at risk on Fraser Island, where the numbers of dingoes are relatively small because of the large tourism industry killing them off.

Because of what is happening on Fraser Island, organizations such as the Australian Dingo Conservation Association have developed refuges for purebred dingoes to stay in. These shelters isolate the purebreds from other hybrids in order to increase the number of purebred dingoes in the wild. One problem with removing all the purebreds from the wild however is that the hybrids will increasingly become more and more different. They will be breeding with other hybrids and creating new unique genotypes, which wasn't happening when they bred with purebreds. Taking purebreds out of the wild also doesn't help the overall problem at hand, but the refuge is trying its hardest to help with 92 dingoes that stay at the center.

Despite all of the negativity and pessimism about the dingo population, one scientist believes that there is hope for the species. Dr. Guy Ballard believes that with constant monitoring and DNA tests, that we can keep the population of dingoes under control and keep the percent of purebreds high. He uses Limeburner's Creek as an example to the success of purebreds despite the numerous amounts of domestic dogs. "The percentage of purebreds in this area is 90%" says Guy, "that gives us hope for the species as a whole".

Many researchers like Guy have explored the extent of the problem with the dingoes, and many have gone as far as to state many of the consequences on livestock and the dingo population. But at this point, you may be wondering: Isn't this a book about koalas? Why did he just take a whole chapter to talk about dingoes?

To this I say that I am about to explore the deep connection between this problem and the koala species, as well as every other species in Australia. Unlike previous scientists and researchers, I will go into depth about what this dingo problem does to the rest of the animals in Australia. How it affects the fragile ecosystems that enumerate the vast territory, and most importantly, how it affects the koala.

When an animal as important as the dingo is in decline, the whole country of Australia experiences the loss.

Chapter Nine:
The Future of Australian Biodiversity

It would be a shame for a book like this one to only grasp the small problems in the animals and landscape of Australia, without addressing the larger one at hand. The reality is, that all these smaller problems foil into one large-scale problem that many other ecosystems, and countries are going through as well. Moving into the future, technology is significantly taking over just about every industry and replacing things that human workers could normally complete. The world is getting more efficient and replaceable. One example of this is robots working at fast food restaurants, or self-checkout lanes at the grocery store. Machines have been seemingly trumping positions that were thought to belong to humans. As human positions become more and more scarce, there is one thing that machines cannot take over: the natural processes of life.

What do I mean by this? Well, there are vastly too many natural processes to name off to you right now, but I'll give a few examples to give you a bigger picture of what I am talking about. Robots cannot run ecosystems because they rely on a complex food web system with various environmental factors such as storms, rainfall, climate and natural disasters. In every ecosystem, there is a perfect number of producers and consumers in order to keep the system alive and well. Robots have not yet perfected photosynthesis, and that is the basis of a functioning ecosystem. It is where all the energy starts, the lowest trophic level of the system. The nitrogen cycle that is performed by nitrifying and denitrifying bacteria that are able to transform molecules to release free nitrogen into the atmosphere and recycle biomass.

These are just some of the many processes that make up our living earth and keep it together, but there are numerous others. These highly complex systems and processes are something that cannot be replicated because they were all formed over many years of revision and rework. These systems are home to thousands of animals, each with their specific niches, some keystone species and others is top of the pyramid carnivores like the dingo.

Technology, while helping human society with efficiency and power, has struggled to have a positive impact on the ecosystems of the planet and animal habitats. Factories that make clothing and food more efficiently, give off carbon fumes that threaten animal homes. Cars that make travel more efficient release CO_2 that is breaking down the ozone layer causing global warming. Plastic bottles, which makes carrying and drinking water more efficient, is causing waste pollution such as the great pacific garbage patch. Expansion of living houses that makes finding a home more efficient causes deforestation and invasion of animal territories. All things that benefit humanity have a cost on the earth. Some more than others, but every action has an equal and opposite consequence.

The Consequence for Australia

Since this book is mainly about koalas and focuses on Australia, that is where we are going to focus our attention. Keep in mind though, that these concepts apply worldwide, even in your own community, and are not entirely exclusive to the country of Australia.

Australia is home to about 5,696 different animal species. 378 of those are mammal species, 828 bird species, 4000 fish species, 300 lizard species, 140 snake species and 50 marine mammal species. Australia is ranked within the top ten countries for biodiversity, but that number is declining. Biodiversity is where the consequences of technology are hitting the most in Australia. Urbanization on the eastern coast has caused the deforestation and death of food for koalas and other species. Pollution in major cities like Brisbane has also caused many koalas to die or adopt illnesses. Over farming is causing the soil to run dry and is causing the malnourishment and death of many native eucalyptus plants.

Australian biodiversity is so important to protect because of the sheer number of endemic species, species that appear nowhere else on the planet. Australia has more endemic species than 98% of other countries and by far has the most endemic mammal species. Most marsupials in Australia cannot be found in any other part of the world because they have grown up and adapted in Australia. Around 84% of all mammals in Australia cannot be found anywhere else in the world. If this isn't a reason for saving Australian biodiversity, then I don't know what is!

Our record on protecting this biodiversity is not good. Since European settlement, 83 species of higher plants have become extinct, 43 types of animals and 19 types of mammals have become extinct. There are also numerous species that have become threatened since settlement such as 209 species of birds and 54 species of mammals.

Despite having such a large biodiversity, Australia has one of the smallest areas of protected land, only about 60,110,000 ha. This, despite being a large number, is only about 7.8% of all the land in Australia. This is a low number for Australia because there are 27% of countries with higher percentages and they are less developed and have less money than Australia does. Australia and Sri Lanka have about the same population and density, but yet the percent of protected land is almost double that of Australia. Sri Lanka isn't the richest of nations either.

Australia, above all other nations, needs to be protected because of its standing at number one on the list of endemic species. It has more than any other country in the world, even the amazon rainforest. Mammals are the animals that are taking the hardest hit from the environmental issues listed above and throughout this novel. Out of the whole world, Australian mammals are ranked as the 6th largest threatened species country. Out of the 300 species of mammals, over 52 of them are either endangered or threatened species. That is 20% of all the mammals in Australia... a large number that needs to decrease.

This is just a broad view, a look into the problem that Australia is facing. What we are about to do next however, is dive into the deep connections within Australian wildlife and see how these problems are affecting specific species. Many species that are in danger are absolutely crucial to the wellbeing of Australian biodiversity as a whole, and there loss could be a major hit to the country.

The Queensland Koala

"Troubling news out of Queensland" the headline for a Wilderness Society article says. "The koala is being pushed to the brink of extinction". This is a huge claim to be made by the Wilderness Society, but it might not be far from the truth. In Queensland, deforestation of the eucalyptus forests has become an industry. The previous government relaxed rules about land clearing and since then, land clearing rates have just about tripled. 300,000 hectares are being cleared per year, which is about the size of the large city of Brisbane.

More than 400 ecologists from around the world agree that deforestation in Australia is a major threat to all of the endemic threatened species such as the koala, many of which are already publishing warnings and studies. One of the main problems of the public right now is that they cannot see the effects. Often times, the effects of deforestation are staggered and will long-term affect the animals. A short-term effect might not be visible. Land clearing also makes animals less resilient to climate change as there food supply dwindles and they cannot find adequate shade or shelter. Land clearing separates species, fragmenting them throughout the continent. This makes them easier prey for predators because they are scattered and have less protection from trees.

All of these factors have caused a decline of 53% per year in koalas around Queensland. The amount of habitable land left available after deforestation makes it very hard for those numbers to decline. We have already discussed thoroughly, why it is hard for koalas to adapt to deforestation and find new habitats, so there is no need to explain it again. But one thing we do need to know is why it is important that the koala doesn't become extinct.

Koalas don't play the biggest role in the Australian ecosystem because they aren't a keystone species, but that doesn't mean that they aren't vitally important. There are many different factors to predict what animal extinctions would do to an ecosystem. One of those factors is: What does a Koala eat?. We have already discussed this very thoroughly in chapter two, so if you want an in depth refresher, you can flip there again. For short, koalas basically just eat eucalyptus leaves. Koalas, despite eating eucalyptus, also help spread the eucalyptus seeds through their dung. Koalas are the number one spreader of the eucalyptus besides artificial growers that are growing it for oil. Without the koala, there would be significantly less spreading of eucalyptus seeds and therefore, less eucalyptus trees.

The next factor is predators: What eats a Koala?. Koalas are the tops of the food chain, so this question doesn't really apply to them too much, but there are still many predators that will feed on dead koalas such as dingoes, feral cats and foxes. The extinction of the species however, wouldn't cause that big of an ecological impact on these predators because they eat so many other types of prey. Losing one wouldn't be much of an issue for these top predators.

The third factor that applies most to the koala is eco-tourism. Eco-tourism is the relationship between the environment and the financially state of the country. It is basically the idea that animals and environmental features bring more people to the desired country and cause it to gain more revenue and income. In Australia, ecotourism is a big deal. There are over 600 ecotourism operators that employ a total of about 6,500 staff workers. The industry makes an average of 250 million dollars a year. Without the koala, this number would drop a lot, especially since the koala also helps maintain eucalyptus forest population. The koala has become an iconic figure of Australia and without it, Australia would be just another destination with unique cities are architecture.

One scientist outlined the extinction of the koala in a very linear pattern:

No more koalas -> No koala dung-> less forests-> expansion of cities and humans-> less incentive to go to Australia-> decline in ecotourism and beauty of Australia.

The Dangerous Link of the Dingo

Dingoes as noted earlier, are one of the many important carnivorous creatures that roam the Australian ecosystem. We have also seen however, that the purebred dingo is on the decline, facing threats of domestication and inbreeding. This not only poses a problem for the dingo population, but also various other species in Australia through means of a trophic cascade. A trophic cascade is when the abundance, or in this case, decline of a species directly or indirectly affects the amounts and well beings of other species in the same ecosystem and food chain. An example of this would be the drastic virus that wiped out most of the Wildebeests in Gorongosa National Park over a century ago. The population diminished greatly and had catastrophic effects on the whole entire park.

The rinderpest virus, which was infecting and killing wildebeests in Gorongosa was one of the main reasons that many ecologists pushed for its conservation and protection. The decline of the wildebeests had made it so there was much more tall grass in the ecosystem because that is the food of the wildebeests. This may seem like a good thing, but it was in fact, everything but that. The tall grasses caused more frequent and devastating fires that wiped out much of the trees and forests, which were feeding grounds and habitats for numerous other species such as Giraffes. This is a trophic cascade in action, one species, the wildebeest, has affected the survival and food source of the Giraffe and other animals simply by not being as numerous as before.

Trophic cascades can be naturally caused, but are not meant to happen for an ecosystem to be in equilibrium. These events are what caused kelp forests to die off in one part of the Aleutian Islands. The lack of sea otters to eat the sea urchins had created a cascading effect to where the numerous sea urchins ate all the kelp. Trophic cascades have the ability to devastate an ecosystem and they could very well do the same for the Australian one with the dingoes on the decline.

Like we did for the koala, to find out the impact of declining dingoes, we must find out what it eats, and what eats it. Dingoes are on the top of the food pyramid and are of the highest trophic level, so it has no natural predators. It does however; eat a lot of different small rodents, birds and animals. Its niche could be described as a population control for all of Australia's small animals and birds. This is the role of many carnivorous predators at the top of the food chain and it has become known as a type of ecological regulation called top down regulation.

This type of regulation pretty much states that the carnivorous species at the top of the trophic levels control how many of lower trophic level organisms there are by preying on them. This is why populations of lower organisms don't expand exponentially.

So what keeps the number of dingoes, or high trophic level species down? The key is another type of regulation: bottom up regulation. In bottom up regulation, the amount of food available to the predatory species determines how many can thrive and survive. This is why predatory species such as lions and dingoes don't grow to huge numbers despite not having any predators of their own. It takes a lot of food to feed carnivores, and food is a limited resource of an ecosystem.

Bottom up regulation brings us to our second question of the dingo issue: what do they eat? Dingoes eat a little bit of everything and are more like scavenging animals. As opposed to the koala eating only one type of food, this range of diet means that it won't affect one species terribly a lot, but it will affect multiple species in smaller increments. Here is a constructed food web of dingo diets.

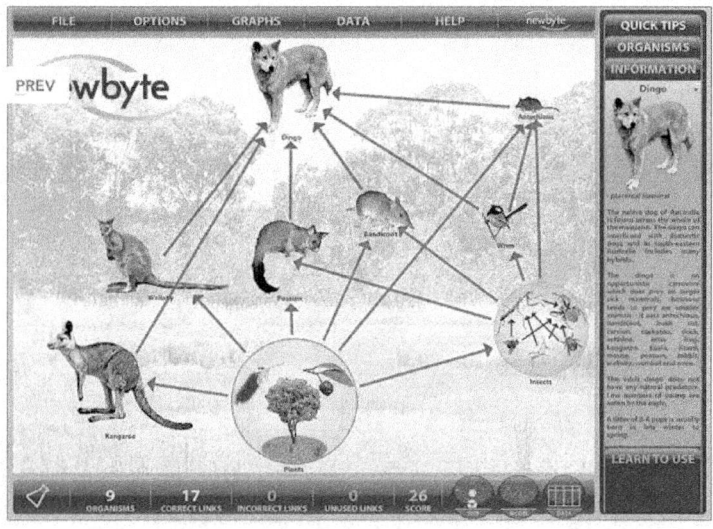

As you can see from the food web, dingoes eat many second level consumers and some third level ones as well. The more levels to a trophic pyramid, the more stable the ecosystem has the potential of being.

Since the dingo is at the top of what seems to be a 4 level pyramid, it is safe to say that the impact of the dingo decline won't be something that is noticed immediately like the wildebeests or sea otters, but rather a gradual trend.

Based on the dingo interbreeding research, it seems that there are two different outcomes for the dingoes that are becoming hybrids. Some are becoming fiercer and requiring more food, while some are becoming more domesticated and ditching their diet all together. We will take a look at the potential outcome or trend of each result from the hybridization.

If the dingo were to become larger from interbreeding and therefore require more food, the ecosystem would probably suffer the biggest hit. The bigger dingoes with bigger diets would eat more animals, overthrowing the top down control mechanism that is in place already. Eating more mammals and birds, according to the food web, would cause an increase in insects and plants because there would be fewer consumers to eat them. The increase of plants and insects, especially insects, could really mess with the balance of the trophic pyramid. The numerous insects could eat too many plants, causing there to be less food for other mammals and animals. This increase in plants too, would facilitate a very large population of insects and cause them to be the enumerating factors of the pyramid. Australia is home to some of the most deadly types of insects such as the Sydney Funnel Web Spider and the European Honey bee. This increase in deadly animals would theoretically decrease ecotourism as well, because who wants to visit an insect oasis.

The second outcome of the dingo decline is that dingoes become domesticated and eat less. This would be abandoning their special niche, which is keeping the populations of small rodents and possums from overpopulating Australia. An increase in these creatures would create a decrease in plants and food availability, which would cause more competition for food and resources. This competition has been seen in the past, and what usually happens is that one species emerges as the best fit to get that food and will become an invasive species or dominant in an ecosystem. This could happen with the lessening amount of food available and would most likely lessen biodiversity in Australia.

Both outcomes would be undesired for the biodiversity of Australia, which is why there are concerns and steps being taken to save the purebred dingo such as creating sanctuaries and breeding areas for them. Many probably don't realize the problem in letting these species interbreed, but those people aren't trained ecologists and wildlife experts. The survival of the dingo is a necessity, as is any other carnivore on the top of a trophic pyramid. In Gorongosa National Park, when the lion population was declining because of lack of wildebeests due to rinderpest, there were immediate efforts to restore that population. Gorongosa had been devastated by the loss of lions, and much of the plants and flora had been destroyed by too many herbivores or war.

Biodiversity as an Importance

Biodiversity is one of the most important factors in the world. It is what makes this planet both sustainable and able to survive catastrophe. If an ecosystem isn't diverse and a virus infects one member of the population of species, or war or fires destroy the food source, that ecosystem will cease to exist. Having a limited number of species in the ecosystem causes that ecosystem to experience more extreme changes in response to environmental stimuli such as global warming or natural disaster. Even though the removal of wildebeests hurt the lion population in Gorongosa, it didn't entirely destroy it because there were other types of food for the lions such as gazelles and zebras. If that ecosystem were to be entirely wildebeests and lions, then the lion species may not fare the same result and recovery as it did.

Biodiversity is also an indication that an ecosystem is doing well and is well maintained. A Shannon-Weiner diversity index, which measures diversity and evenness, which is how many of each species there are in comparison to other species, is used to see how well an ecosystem is doing. This number helps ecologists determine if ecosystems are doing ok, and there are many ways that they gather their data. One way that any citizen scientist can contribute to this research is through trail cams. Cameras have been set up in various recovering ecosystems such as Gorongosa and Serengetti National Parks. Anyone can go online to spot animals on these cameras and help the various organizations to gather data for measuring biodiversity. Having a community backing up these organizations causes them to work much more efficiently and build awareness for the issues.

But most importantly regarding this problem from a human perspective is our desire and need for biodiversity. I could go through every natural explanation as to why biodiversity is important, but if it is not rationalized and applied to humanity, then few will sympathize with the cause. Humans, whether they realize it or not, have a natural affinity or focus on life over non-life. This is why we own pets despite them not meeting our basic needs or providing us with any material benefit. This is also why we keep plants in our backyard, away from any other human eye. These things seem quite useless without the explanation of phenomena called Biophilia.

Biophilia, a hypothesis that has been popularized by E. O. Wilson, is defined as "the innate tendency to focus on life and lifelike processes". This drive to be with nature and the living world is a combination of both our genetic makeup and an unexplainable force inside every person that wants to be with nature. This force that causes us to focus on life rather than non life comes from culture and our adaptation to this environment that is filled with life. This hypothesis also can explain space expeditions, as most of them are trying to find life on other planets in order that we can study it and find out if it can sustain us. Most scientists would be much more thrilled to find microorganisms living in the soil of Mars, than to find a rare rock on Mercury.

Biophilia, although a not very easy to falsify hypothesis, has been tested in surgery recovery centers. One group of patients was given a view of a forest, while one group was given a view of a brick wall outside their window. The group with the forest view was reported to take half as long to recover and took three times less pain medications than the group looking at the wall. Based on this study, it seems that Biophilia has a potential to change our behavior and wellbeing as a species.

From a human perspective, Biophilia is the reason that we should focus on saving and maintaining biodiversity. Humans should all have a natural tendency to focus on life, and if that life is missing from their lives, then behavior could change or people would feel like they are lacking in something. It is one of those effects that you do not realize the consequences until it happens. We will not realize our natural tendency for life until we see the absence of it. Sticking a person in a barren desert with only the necessities to live would be an example of this, as the person, despite being alive, would be lacking in something. But do we really want to reach a point of nonlife, to realize that we in fact do have a tendency to focus on life? Absolutely not. Because at that point, it will be too late to reverse the effects.

The basic summary of why we need biodiversity and ecosystems from a common human perspective is that without them, humans would feel a dire lack of something in their life, and that something would be extremely hard to get back or recreate.

Whether you look at it from a biological perspective, or a normal human perspective, ecosystems and biodiversity is worth saving because it is one of the building blocks and foundations of the Earth. It is like removing a cornerstone from a building, or a keystone from a bridge: it would cause catastrophic disaster. Australia, home to 5,696 different animal species, is worth protecting. Whether you care about the matter or not, I hope that by reading this you have become aware of your innate desire towards life over non life, preservation over destruction, forest over wasteland and most of all: biodiversity over a world in which every ecosystem is a concrete jungle.

Chapter Ten:

A Case for Life

The journey of writing this book has brought me to the place I am today. This arduous path of research, logic and reasoning, and writing has caused my understanding of animals and the natural world to increase vastly. Remembering back to when I first visited the zoo, unaware that those animals were mere representations of the wild species as a whole, I now realize that this has opened my eyes to the world of the wild: The world where animals hunt and kill each other and where it is a battleground for survival. If you watch the Planet Earth series by BBC, you will see that although the planet is marvelous, there are always competitions for life and death.

Constantly pitted against each other in the ring for survival are predator and prey. Can the prey outrun the predator? Will the predator eventually give up? These could be questions asked in a normal natural conflict, but is this truly the case with koalas? Koalas face a different danger, one that isn't natural. With no natural predators, they face against the hardest competition of them all: mankind.

The good thing about mankind however, is that unlike wild animals and natural predators, mankind has a voice. It isn't natural for man to kill animals; it is actually something that doesn't fit into the rules and theories of the universe such as natural selection and evolution. The original intent for humanity was to be stewards over the earth, not take part in the brutal cycle of natural life. In order to be the stewards we were originally meant to be however, we must learn the basic principles that run the natural world, and how we can help keep them alive and well.

Learning the principles of ecology, ecosystems and habitats is the basis for this human level of self-awareness needed to help animals and sustain this world. If you know that destroying trees will certainly kill an ecosystem and therefore hurt ecotourism and biodiversity, then you are less likely to destroy.

But not destroying isn't quite enough in the case of the koala. We have destroyed too much already, and now must focus on restoring. These two types of preservation are the essential ingredients to a healthy planet and sustainable world. But how, you may ask, do we restore the koala population? Luckily for you, there are already people trying to solve that problem like me that are here to guide you.

The main association for saving koalas, and one that I have had the opportunity to partner with through my koala music, is the Australian Koala Foundation or AKF for short. This organization is responsible for the majority of restoration and preservation of koalas in Australia. They have pushed major legislation out in the government and are trying to push through a Koala protection act. The AKF does everything from planting new trees in the forests, to saving koalas and putting them in shelters or reserves.

With all this information, one may stand on the sidelines because it is the easy more taken path. You can be just like an average zoo attendee, looking and admiring on the work of others and the beauty of nature. We live in a world where this option is often taken. For example, when somebody drops their books during school, how many people just walk by and ignore it. They know fully well how to help, but don't because it is out of their way or takes too much time. After reading this book you will have all the information you need to make your case in defense of the koalas. You will know of all the threats of climate change, and why koalas cannot defend themselves. You will know of the growing deforestation and starvation problems that they face. You will know of all the human urbanization causing them to find new homes. On top of this knowledge, you will also know why koalas are worth saving. You will know that the ecosystems in Australia and biodiversity will take a plummet. You will know that koalas are one of the most unique and interesting animals on the planet. You will know that koalas are one of the few rule-breaking animals we can study and make points against famous hypotheses and theories. You will also know that koalas are worth saving because of the potential research in the field of behavior and behavioral evolution. You have all the information to make your case... will you make it?

I know that for me, this information was enough for me to make mine. I turned from being an average zoo attendee, to being a worker at the zoo, trying to improve the lives of the animals. I turned from being an average high school student, to an author of one of the most extensive koala novels to date. I've certainly found the motivation to be a restorer to this world and to help preserve it because I know the facts, the scientific consensus, and the damages.

But anyone can read about how much we are losing on this planet, how many species have gone extinct or are endangered. What I am trying to do is make a case for life, one that promotes action and requires effort to perform. If we treated animals and the animal kingdom as they deserve, human inhumanity would stand out all the more appallingly. We could turn from a human society to a *humane* society for once in our history.

All life is one. We are connected to this world more than we know it. The same atoms that make up our bodies and flow through us run through animals as well. We have an innate tendency to focus on life. We need animals to make us complete.

"Anyone who studies a wild animal faces the challenge of, in effect, making a case for its life on earth. I pray that mine is strong enough." - Alexandra Morton, Listening to Whales.

References

American Museum of Natural History. "Polar Bear Changes Diet." *ScienceDaily*. ScienceDaily, 22 Jan. 2014. Web.

American Scientist. N.p., n.d. Web.

"Australia's Biodiversity - A Summary." *The Wilderness Society*. N.p., 10 Dec. 2015. Web.

Bournemouth University. "Climate Change: Animals Need Time To Adapt To New Habitats And Survive." *ScienceDaily*. ScienceDaily, 2 Aug. 2009. Web.

Brady, Heather. "Why Koalas Are Suddenly Drinking Extra Water." *National Geographic*. National Geographic Society, 16 Aug. 2017. Web.

Bryner, Jeanna. "Evolution Can't Go Backward." *LiveScience*. Purch, 23 Sept. 2009. Web.

"Cracking the Koala Code." *PBS*. Public Broadcasting Service, 27 Oct. 2014. Web.

Doran, John. "Soil Health and Sustainability: Managing the Biotic Component of Soil Quality." *Digital Commons*. N.p., n.d. Web.

Ellis, Bill. "Urban Koala Update." *San Diego Zoo Blog*. N.p., 25 Jan. 2012. Web.

"Environment News." *ABC News*. Australian Broadcasting Corporation, n.d. Web.

"Feature Article - Australia's Biodiversity (Feature Article)." *Australian Bureau of Statistics, Australian Government*. N.p., n.d. Web.

"Feature Article - The Soils of Australia (Feature Article)." *Australian Bureau of Statistics, Australian Government*. N.p., n.d. Web.

"Genetically Engineered Eucalyptus Plantations Threaten US South." *WilderUtopia.com*. N.p., 08 Nov. 2014. Web.

Ghose, Tia. "Why Koalas Hug Trees." *LiveScience*. Purch, 03 June 2014. Web.

"The Invasion of Genetically-Engineered Eucalyptus." *Organic Consumers Association*. N.p., n.d. Web.

Jun, Yang Suk, Purum Kang, Sun Seek Min, Jeong-Min Lee, Hyo-Keun Kim, and Geun Hee Seol. "Effect of Eucalyptus Oil Inhalation on Pain and Inflammatory Responses after Total Knee Replacement: A Randomized Clinical Trial." *Evidence-based Complementary and Alternative Medicine : ECAM*. Hindawi Publishing Corporation, 2013. Web.

"Koalainfo.com." *Koala Info*. N.p., n.d. Web.

Langly, Liz. "Do Animals Dream?" *National Geographic*. National Geographic Society, 08 Sept. 2015. Web.

Marinacci, Peter. "Wombat Behavior." *Wombania*. Wombania, 03 Dec. 2014. Web.

Mcallister, Peter. "Genetically Pure Dingoes Face Extinction." *Australian Geographic*. N.p., 11 Mar. 2011. Web.

Morton, Alexandra. *Listening to Whales: What the Orcas Have Taught Us*. N.p.: Ballantine, 2004. Print.

Musser, Anne. "Australian Museum." *Nimiokoala Greystanesi - Australian Museum*. N.p., 15 July 2009. Web.

Neena Bhandari in Sydney for IPS, Part of the Guardian Environment Network. "Climate Change Compounds Rising Threats to Koala." *The Guardian*. Guardian News and Media, 30 Apr. 2013. Web.

Safina, Carl. *Beyond Words: What Animals Think and Feel.* N.p.: Picador, 2016. Print.

Trafton, Anne. "Evolution, Reversed." *Research & Development.* N.p., 14 May 2015. Web.

Turner, Vivienne. "Banksia Pollen as a Source of Protein in the Diet of Two Australian Marsupials Cercartetus Nanus and Tarsipes Rostratus." *Oikos* 43.1 (1984): 53-61. *JSTOR.* Web.

University of Guelph. "Captive Animals Show Signs of Boredom, Study Finds." *ScienceDaily.* ScienceDaily, 14 Nov. 2012. Web.

Vanessa Wong Lecturer in Soil and Land Management, School of Geography and Environmental Science, Monash University, and Robert Edis Soil Scientist, University of Melbourne. "A More Sustainable Australia: We Need to Talk about Our Soils." *The Conversation.* N.p., 13 Aug. 2017. Web.

Veterinärmedizinische Universität Wien. "Scent Marking - the Mammalian Equivalent of Showy Plumage." *ScienceDaily*. ScienceDaily, 31 Oct. 2013. Web.

Vitelli, Romeo. "What Can Solitary Mammals Teach Us About Autism?" *Psychology Today*. Sussex Publishers, 02 June 2014. Web.

www.ingramcontent.com/pod-product-compliance
Lightning Source LLC
Chambersburg PA
CBHW070245230526
45470CB00002B/487